SpringerBriefs in Applied Sciences and Technology

Forensic and Medical Bioinformatics

Series editors

Amit Kumar, Hyderabad, India
Allam Appa Rao, Hyderabad, India

More information about this series at http://www.springer.com/series/11910

Bita Mokhlesabadifarahani
Vinit Kumar Gunjan

EMG Signals Characterization in Three States of Contraction by Fuzzy Network and Feature Extraction

 Springer

Bita Mokhlesabadifarahani
Biomedical Engineering Department
Amirkabir University of Technology
Tehran
Iran

Vinit Kumar Gunjan
Annamacharya Institute of Technology
  and Sciences (AITS)
Rajampet
India

ISSN 2196-8845   ISSN 2196-8853   (electronic)
SpringerBriefs in Applied Sciences and Technology
ISBN 978-981-287-319-4   ISBN 978-981-287-320-0   (eBook)
DOI 10.1007/978-981-287-320-0

Library of Congress Control Number: 2014958016

Springer Singapore Heidelberg New York Dordrecht London

Printed on acid-free paper

Springer Science+Business Media Singapore Pte Ltd. is part of Springer Science+Business Media
(www.springer.com)

This study is dedicated to my parents; my mother, Sakineh Shirzadimokhles, and my father, Aboulfazl Rabieenejad, both of whom gave me the foundation of value, honesty and commitment towards work.

Ever since then, I have been able to appreciate the value of reading and lifelong learning.

Foreword

This book is an enthusiastic contribution containing one of the best research works in the field of EMG Signal Processing.

Still another element is provided by many interesting data on signal feature extraction with fuzzy network and an abundance of colourful illustrations. On top of that, there are innumerable historical vignettes that interweave fuzzy network and EMG signal processing in a very appealing way.

Although the emphasis of this work is on Signal Processing, it contains much—indeed to anyone with a fascination with the world of molecules and neural networks. The authors have selected a good number of prominent molecules as the key subjects of their essays. Although these represent only a small sample of the world of biologically related molecules and their impact on our health, they amply illustrate the importance of this field of science to humankind and the way in which the field has evolved.

I think that the contributors can be confident that there will be many grateful readers who will have gained a broader perspective of the disciplines of signal processing disorders and their remedies as a result of their efforts.

Hyderabad, India   Bita Mokhlesabadifarahani
                                                               Vinit Kumar Gunjan

Preface

This book presents research-based practices which are outcomes of my experiments on EMG Technique and Feature Extraction. This piece of work can be utilized for reading, teaching, and research purposes.

Chapter 1—talks about EMG Technique and Feature Extraction. The EMG signal which is electrical indication of the neuromuscular actuation connected with a contracting muscle.

Chapter 2—focuses on methodology used for working with *EMG Dataset.*

Chapter 3—is about results of the EMG signal processing executed in experiments where after data acquisition phase all recorded signals underwent noise filtering as preprocessing phase.

The last chap. 4 records conclusions and inferences of the research work.

Acknowledgments

I wish to extend my appreciation and thanks to my professor Dr. Ali Fallah and Dr. Ali Maleki at the Amirkabir University of Technology, Tehran, Iran, you always prioritized my needs.

My Mother and Father, you are merciful and through you my God everything is possible.

My dear friend Hossein Roshandel your support, advice, and motivation made this study a success. Sometimes conditions were not favorable to you, but you always offered services.

To my sister, Toktam Mokhlesabadi, I acknowledge your support; and Mellika, you have been patient, motivational, and supportive, I will always cherish.

Contents

Overview of the Book

Neuromuscular, musculoskeletal disorders and injuries highly affect the lifestyle and the motion abilities of an individual. The primary purpose of this work is to develop a systematic method to detect the level of muscle power declining in musculoskeletal and neuromuscular disorders. To this aim, EMG signals of five skeletal muscles as biceps, deltoid, triceps, tibialis anterior, and quadriceps muscles are recorded in three states of isometric contraction (ISO), maximum voluntary contraction (MVC), and dynamic contraction from 22 normal subjects aged between 20 and 30; half of them are male. Totally, 14 combinatory extracted features are analyzed to find which of them or a combinatory set of them are discriminative and selective for muscle force quantification and classification. The neuro-fuzzy system is trained with 70 % of the recorded EMG cut off windows and then employed for classification and modeling purposes. For each muscle the most effective extracted features are found for males and females separately by a reference classifier. In the experiments, after the optimum set of combinatory features is found by a reference classifier, the neuro-fuzzy classifier is validated in comparison to other well-known classifiers in classification of the recorded EMG signals with the three states of contractions corresponding to the extracted features. Then, different structures of the neuro-fuzzy classifier are also comparatively analyzed to find the optimum structure of the classifier used.

Keywords Biceps · Deltoid · Triceps · Tibialis anterior · Quadriceps · Isometric contraction · Maximum voluntary contraction · Dynamic contraction · EMG signal characterization · Neuro-Fuzzy classifier

Chapter 1
Introduction to EMG Technique and Feature Extraction

The electromyography (EMG) signal is electrical indication of the neuromuscular actuation connected with a contracting muscle. It is an exceedingly complicated sign which is influenced by the anatomical and physiological properties of muscles, the control plan of the fringe sensory system, and also the attributes of the instrumentation that is utilized to identify and watch it. Most of the connections between the EMG signal and the properties of a contracting muscle which will be quickly utilized have developed serendipitously. The absence of a fitting portrayal of the EMG signal is likely the most noteworthy single variable which has hampered the improvement of EMG into an exact discipline.

This section will show two fundamental ideas. The first is a discourse of an organized approach for deciphering the data content of the EMG signal. The scientific model which is created is built with respect to current learning of the properties of contracting human muscles. The degree to which the model helps to the understanding of the sign is confined to the constrained sum of physiological learning right now accessible. On the other hand, even in its available structure, the demonstrating methodology supplies an edifying knowledge into the arrangement of the EMG signal.

EMG is the investigation of muscle capacity through examination of the electrical signs radiated amid brawny constrictions. EMG is frequently misused and abused by numerous clinicians and scientists. Ordinarily even accomplished electromyographers neglect to give enough data and detail on the conventions, recording supplies and techniques used to permit different analysts to reliably recreate their studies. Assuredly, this part will elucidate some of these issues and give the peruser a premise for having the capacity to direct EMG examines as a feature of their on-going exploration.

EMG is measuring the electrical sign connected with the enactment of the muscle. This may be willful or automatic muscle compression. The EMG action of willful muscle withdrawals is identified with pressure. The practical unit of the muscle compression is an engine unit, which is embodied a solitary alpha

© The Author(s) 2015
B. Mokhlesabadifarahani and V.K. Gunjan, *EMG Signals Characterization in Three States of Contraction by Fuzzy Network and Feature Extraction*, Forensic and Medical Bioinformatics, DOI 10.1007/978-981-287-320-0_1

engine neuron and all the strands it exhausts. This muscle fiber contracts when the activity possibilities (drive) of the engine nerve which supplies it achieves a depolarization limit. The depolarization creates an electromagnetic field and the potential is measured as a voltage. The depolarization, which spreads along the film of the muscle, is a muscle activity potential. The engine unit activity potential is the spatio and transient summation of the individual muscle activity possibilities for all the filaments of a solitary engine unit. In this way, the EMG sign is the arithmetical summation of the engine unit activity possibilities inside the pick-up territory of the terminal being utilized. The pick-up zone of a cathode will quite often incorporate more than one engine unit on the grounds that muscle filaments of distinctive engine units are mixed all through the whole muscle. Any bit of the muscle may contain filaments having a place with upwards of 20–50 engine units.

A solitary engine unit can have 3–2,000 muscle filaments. Muscles controlling fine developments have more modest quantities of muscle filaments for every engine units (typically short of what 10 strands for every engine unit) than muscles controlling extensive terrible developments (100–1,000 strands for every engine unit). There is a chain of importance game plan amid a muscle compression as engine units with less muscle strands are normally enlisted initially, taken after by the engine units with bigger muscle filaments. The quantity of engine units for every muscle is variable all through the body.

With the end goal of this part there are two fundamental sorts of EMG: clinical (frequently called indicative EMG) and kinesiological. Symptomatic EMG, normally done by physiatrists and neurologists, are investigations of the characteristics of the engine unit activity potential for span and abundancy. These are commonly done to help analytic neuromuscular pathology. They likewise assess the spontaneous releases of loose muscles and have the capacity confine single engine unit action. Kinesiological EMG is the sort most found in the writing in regards to development dissection. This kind of EMG studies the relationship of bulky capacity to development of the body fragments and assesses timing of muscle action with respect to the developments. Moreover, numerous studies endeavor to look at the quality and energy generation of the muscles themselves.

There is a relationship of EMG to numerous biomechanical variables. Regarding isometric withdrawals, there is a positive relationship in the increment of pressure inside the muscle as to the plentifulness of the EMG sign recorded. There is a slack time, on the other hand, as the EMG adequacy does not specifically match the manufacture up of isometric pressure. One must be watchful when attempting to gauge energy creation from the EMG signal, as there is sketchy legitimacy of the relationship of power to plentifulness when numerous muscles are intersection the same joint, or when muscles cross different joints. At the point when taking a gander at muscle movement, as to concentric and unconventional compressions, one finds that flighty withdrawals create less muscle action than concentric withdrawal when conflicting with equivalent energy. As the muscle exhausts, one sees a diminished strain regardless of steady or considerably bigger adequacy of the muscle action. There is a loss of the high-recurrence segment of the sign as one uniform, which can be seen by a diminishing in the average recurrence of the muscle signal. Amid movement, there has a tendency to be an association with EMG and speed of the development. There is an opposite relationship of quality creation with concentric withdrawals and the velocity of development, while there is a positive relationship of quality generation with unconventional constrictions and the pace of development. One can deal with even more a heap with unconventional contractions at higher rate. For instance: If a weight was huge and you brought it down to the ground in a quick, yet controlled way, you took care of a vast weight at a rapid through flighty compressions. You would not have the capacity to raise the weight (concentric withdrawal) at the pace you had the capacity lower it. The constrained generation by the strands are not so much any more noteworthy, yet you had the capacity handle a bigger measure of weight and the EMG action of the muscles taking care of that weight would be littler. In this manner, we have a reverse relationship for concentric withdrawals and positive relationship for offbeat constrictions regarding pace of development.

As to recording the EMG signal, the adequacy of the engine unit activity potential relies on upon numerous elements which include: breadth of the muscle fiber, separate between dynamic muscle fiber and the location site (fat tissue thickness), and sifting properties of the terminals themselves. The target is to get a sign free of noise (ie., development relic, 60 Hz ancient rarity, and so on). Along these lines, the anode sort and enhancer qualities assume a vital part in getting a commotion free flag.

Development and position of appendages are controlled by electrical signs going here and there and then here again between the muscles and the fringe and focal sensory system. At the point when pathologic conditions emerge in the engine framework, whether in the spinal rope, the engine neurons, the muscle, or the neuromuscular intersections, the attributes of the electrical flag in the muscle change. Watchful enlistment and investigation of electrical flag in muscle (electromyograms) can along these lines be an important support in finding and diagnosing abnormalities in the muscles as well as in the engine framework overall. EMG is the enlistment and elucidation of these muscle activity possibilities. As of not long ago, electromyograms were recorded fundamentally for exploratory or analytic purposes; be that as it may, with the headway of bioelectric innovation, electromyograms likewise have turned into an essential device in accomplishing manufactured control of appendage development, i.e., practical electrical incitement (FES) and restoration. This part will concentrate on the symptomatic application of electromyograms. Since the ascent of advanced clinical EMG, the specialized strategies utilized as a part of recording and dissecting electromyograms have been managed by the accessible innovation. The concentric needle cathode introduced by Adrian and Bronk in 1929 gave a simple-to-utilize anode with high mechanical qualities and steady, reproducible estimations. Supplanting of galvanometers with high-pickup speakers permitted more modest terminals with higher impedances to be utilized and possibilities of littler amplitudes to be recorded. With these specialized accomplishments, clinical EMG soon advanced into a very particular field where electromyographists with numerous years of experience read and deciphered long paper EMG records focused around the visual appearance of the electromyograms. Gradually, a more quantitative methodology developed, where peculiarities, for example, potential term, crest-to-top adequacy, and number of stages, were measured on the paper records and contrasted, and a set of ordinary information accumulated from solid subjects of all ages. In the most recent decade, the universally useful rack-mounted supplies of the past have been supplanted by thus nomically planned EMG units with incorporated machines. Electromyograms are digitized, transformed, put away on removable media, and shown on machine screens with screen designs that change in agreement with the sort of recording and dissection picked by the examiner.

In light of this, this section gives an acquaintance with the essential ideas of clinical EMG, a survey of fundamental life structures, the source of the electromyogram, and a percentage of the primary recording strategies and sign investigation methods being used.

1.1 Structure

Muscles represent around 40 % of the human mass, running from the little extraocular muscles that turn the eyeball in its attachment to the expansive append- age muscles that create motion and control carriage. The configuration of muscles differs relying upon the scope of movement and the energy pushed. In the most straightforward game plan (fusiform), parallel filaments expand the full length of the muscle and connect to tendons at both closures. Muscles creating an extensive power have a more entangled structure in which a lot of people short muscle fila- ments join to a level tendon that stretches out over an expansive part of the muscle. This mastermind ment (unipennate) expands the cross-sectional territory and along these lines the contractile energy of the muscle. At the point when muscle filaments fan out from both sides of the tendon, the muscle structure is alluded to as bipennate.

A lipid bilayer (sarcolemma) encases the muscle fiber and differentiates the intracellular myoplasma from the interstitial liquid. Between neighboring fila- ments runs a layer of connective tissue, the endomysium, made principally out of collagen and elastin. Packs of filaments, fascicles, are held together by a thicker layer of connective-tissue called the perimysium. The entire muscle is wrapped in a layer of connective tissue called the epimysium. The connective tissue is constant with the tendons appending the muscle to the skeleton.

In the myoplasma, meager and thick fibers interdigitate and structure short, seri- ally associated indistinguishable units called sarcomeres. Various sarcomeres associ- ate end to end, accordingly framing longitudinal strands of myofibrils that augment the whole length of the muscle fiber. The aggregate shortening of a muscle amid constriction is the net impact of all sarcomeres shortening in arrangement all the while. The individual sarcomeres abbreviate by structuring cross-connects between the thick and dainty fibers. The cross-scaffolds pull the fibers to one another, along these lines expanding the measure of longitudinal cover between the thick and slight fibers. The thick grid of myofibrils is held set up by a structural system of between intercede fibers made out of desmin, vimetin, and synemin (squire, 1986).

At the site of the neuromuscular intersection, each one engine neuron structures insurance grows and innervates a few muscle strands circulated practically equally inside a curved or roundabout district extending from 2 to 10 mm in width. The engine neuron and the muscle strands it innervates constitute a useful unit, the engine unit. The cross-area of muscle involved by an engine unit is known as the engine unit domain (MUT). A regular muscle fiber is just innervated at a solitary point, found inside a cross-sectional band alluded to as the end-plate zone. While the width of the end-plate zone is just a couple of millimeters, the zone itself may stretch out over a huge piece of the muscle. The quantity of muscle strands for every engine neuron (i.e., the innervation degree) ranges from 3:1 in extraneous eye muscles where fine-evaluated withdrawal is obliged to 120:1 in some appendage muscles with coarse development (kimura, 1981). The filaments of one engine unit are intermixed with strands of other engine units; hence, a few engine units dwell inside a given cross-area. The filaments of the same.

EMG is technique to evaluate and record the electrical activity of the muscle and is a valuable device to assess neuromuscular disorders. Computer-aided EMG has evolved as an indispensable tool in the everyday activity of neurophysiology laboratories in facilitating quantitative analysis and decision making in the clinical neurophysiology, rehabilitation, sport medicine and human physiology. EMG findings are used to detect and describe special disease processes affecting the Motor Unit (MU), which is the smallest functional unit of the muscle [1] (Fig. 1.1).

In EMG-based model recognition, sEMG signal is pre-processed from the spectral frequency component of the signal and is extracted with some features before performing classification [2] (Fig. 1.2).

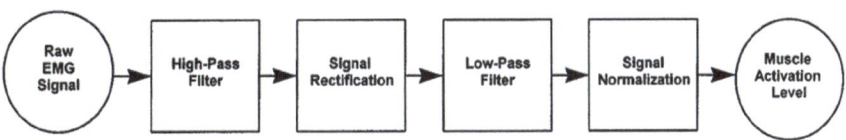

Fig. 1.1 Electromyogram signal processing algorithms to estimate the level of muscle activity

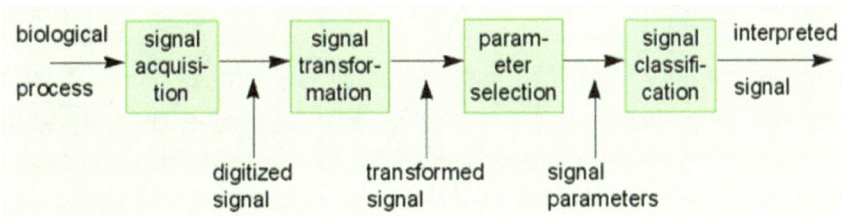

Fig. 1.2 Signal processing

Normally, in pre-processing and signal condition process, procedure to remove noise is a significant step to reduce noises and improve some spectral component part [3]. Next important step, feature extraction, is used for highlighting the relevant structures in the sEMG signal and rejecting the noise and unimportant sEMG signal [4]. The success of EMG pattern recognition depends on the selection of features that represent raw sEMG signal for its classification. This study is enforced by the fact that the limitation of the solution to remove WGN in the pre-processing step and EMG-based gestures classification need to conduct the extraction step. The selection of the feature that tolerance of WGN and modification of existing EMG feature to improve the robust property are proposed. Resultantly, WGN removal algorithms in the preprocessing step are not needed.

Feature extraction is a method to extract the useful information that is hidden in surface EMG signal and to remove the unwanted EMG parts and interferences [4, 5] (Fig. 1.3).

Some features are strong across different kinds of noises; consequently, intensive data pre-processing methods shall be avoided to be implemented [6]. In addition, appropriate features approaches high classification accuracy [7]. Three properties have been suggested for use in quantitative comparison of their capabilities that include maximum class separability, robustness, and complexity [4, 5]. Although many research works have mainly tried to explore and examine an appropriate feature vector for numerous specific EMG signal classification applications (e.g. [4–8]), there are other works which made deeply quantitative comparisons of their qualities, particularly in redundancy point of view [9].

Furthermore, most recent EMG signal classification studies have still employed set of feature vectors that carried a number of redundant features (e.g. [10–17]).

In 1975, the Graupe and Cine showed that a fourth-order time-series model of EMG signals can be classified by a linear discrimination function [18], but this technique involves a high complexity in computation. The results of Kelly and Parker's work illustrated that a Hopfield neural network could produce AR coefficients from EMG signals in a shorter time [19]. Furthermore, Saridis and Gootee presented integral absolute value and zero-crossing features that could produce appropriate feature space in turn to classify arm motions [20]. Zardoshti and folks [4] extracted few features such as integral of absolute value, variance,

Fig. 1.3 For feature extraction template+

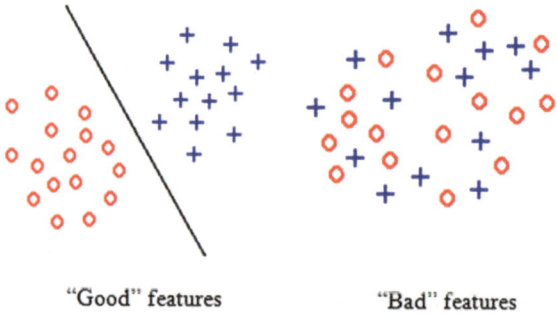

"Good" features   "Bad" features

number of zero crossings and auto-regressive model parameters from upper limb EMG signals and then evaluated them with K-nearest neighbor (a non-parametric classifier). They presented a new feature, EMG histogram, which is highly suitable for the classification of hand motions, and showed that this feature is appropriate to calculate both speed and noise tolerance. Chang and folks [21] used the variance of the rectified wave envelope and IAV features and Mahalonobius distance to classify four pre-shaping grasp movements. They also showed that these features could classify movements up to 90 % accuracy. Kang and folks (1995) compared AR and cepstrum coefficients and showed that the cepstrum coefficients are quite useful to improve classification rate. The time frequency transform has also been introduced as a new mathematical approach to time–frequency domain. Biomedical signals, especially EMG signals, have been processed by time–frequency transforms in order to extract more representative features to improve rate of classification of motions. In this way, Jung and folks [22] imposed Wigner–Ville transform on upper limb EMG signals to classify six different movements. Wellig and Moschytz [23] too used packet wavelet transform to decompose EMG signals and reduce misclassification rate. Liyu et al. [24] distinguished four forearm motions by decomposing two channels of EMG signals with wavelet transform in six levels, and finally classified these coefficients by an artificial neural network (ANN) classifier. Abel and folks [25] by applying inter-scale local maximum method on wavelet coefficients of EMG signals presented new features, which improved classification rate among neuropathic, myopathic and normal groups. Englehart et al. [26] extracted upper limb EMG signals from four channels and then, by extracting wavelet coefficients, reduced their dimensions by PCA transform, and finally misclassification rate was decreased. Although literature includes many papers which explore extraction of features from EMG for controlling prosthetic limbs, there have been few works which make quantitative comparison of their quality. Christodoulos and Pattichis [27] used an ANN based on unsupervised learning and a statistical pattern recognition technique based on Euclidean distance to analyze a total of 1213 MUAP's obtained from 12 normal subjects, 13 subjects suffering from myopathy, and 15 subjects suffering from motor neuron disease and they reported success rate for used ANN technique as 97.6 % and for statistical technique 95.3 %. In 2005, Huang et al. [41] used Gaussian Mixture Model on a 12 subject database to classify subjects according to feature sets including time-domain (TD) features and autoregressive features with root mean square value (AR + RMS). They reported Gaussian mixture model (GMM) achieves 96.91 % classification accuracy using a AR + RMS + TD feature set and attains 96.3 % classification accuracy using a AR + RMS feature set for distinguishing six limb motions. Tsenov et al. [28] in 2006 exploited signal recorded at surface of skin of forearm to provide recognition of movement of hand and finger movements of healthy subjects. They utilized radial basis function (RBF), multilayer perceptron (MLP) and LVQ networks to classify signals based on time domain extracted features as Mean Absolute Value, Variance, Waveform Length, Norm, Number of Zero Crossings, Absolute Maximum, Absolute Minimum, Maximum minus Minimum and Median Value. They reported average classification rates of 92.64 % (10 neurons in hidden layer as best case), 83.82 % (spread value 0.7) and 88.23 % (28 competitive neurons as best case) for MLP, RBF and LVQ networks, respectively.

In 2007, Yoshikawa et al. [29] extracted features MAV, VAR, WL, ZC, Absolute Maximum, Absolute Minimum and Median Value for motion classification based on SVM. In same year, Momen et al. [30] presented a real-time EMG classifier of user-selected intentional movements for signals recorded from forearm extensor and flexor muscles of seven able bodies-subjects and one congenital amputee. Segmentation of feature space was performed using fuzzy C-means clustering. It was reported that with only 2 min of training data from each user classifier discriminated four different movements with an average accuracy of 92.7 % $\pm$ 3.2 %. It was stated in their work that presented method may facilitate development of dynamic upper extremity prosthesis control strategies using arbitrary, user-preferred muscle contractions (Fig. 1.4).

Hudgins et al. [31] were pioneers in developing a real-time pattern-recognition-based MCS. Using TD features and a MLP neural network, they succeeded in classifying four types of upper limb motion, with an accuracy of approximately 90 %. This work was continued over last 15 years, by employing various classifiers, such as linear discriminant analysis (LDA) [32, 33], MLP/RBF neural networks [34], time-delayed ANN [35], fuzzy [36, 37], Neuro-Fuzzy [38], fuzzy ARTMAP networks [39], fuzzy-MINMAX networks [40], GMMs [41–43], and hidden Markov models (HMMs) [44]. Vuskovic and Du [39] introduced a modified version of a fuzzy ARTMAP network to classify prehensile MESs. Englehart et al. [32] showed that LDA, outperforms MLP on time-scale features that are dimensionally reduced by PCA. In addition, significant results were achieved using probabilistic approaches. Chan and Englehart [44] applied an HMM to discriminate six classes of limb movement based on a four-channel MES. It resulted in an average accuracy of 94.63 %, which exceeded an MLP-based classifier used in [33] (93.27 %). Furthermore, Huang et al. [41] and Fukuda et al. [42] developed a GMM as a classifier in their MCS; former showed an accuracy of approximately 97 %. Englehart et al. [33] introduced a continuous classification scheme that provided more robust results for a shortened segment length of signal, and high-speed controllers. Oskoei and Hu [7] employed SVM for classification of upper limb motions suing myoelectric signals. They used another method to adjust SVM parameters before classification, and examined overlapped segmentation and majority voting to improve controller performance. They also used a TD multi-feature set (i.e., MAV + WL + ZC + SSC) as signal features for classification.

Fig. 1.4 EMG and IEMG signals with feature extraction frame

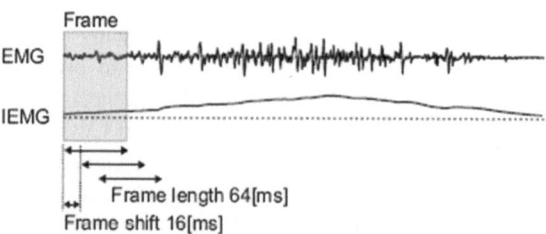

In this book, we follow a high quality EMG feature space which has following properties:

Maximum class separability. A high quality feature space which results in clusters that have maximum separability or minimum overlap. This ensures lowest possible misclassification rate.

Robustness. Lowest possible sensitivity of feature space cluster separability to noise samples.

Complexity. Lowest possible computational complexity of features (and clusters) so that procedure can be implemented with reasonable hardware and in a real-time manner.

Chapter 2
Methodology for Working with EMG Dataset

2.1 EMG Dataset

In our experiments, data acquisition system include PowerLab, 16sp, and Dual BioAmp manufactured by ADInstruments Ltd. and software Chart V5.0 with sampling rate adjusted at 2 kHz, recording signal amplitude 2 mV, primary low-pass filter 1 with cutoff frequency 500 Hz, and primary high-pass filter 2 with cutoff frequency 0.3 Hz. Data are outputted in txt or excel format which are readable in MATLAB for data processing. MATLAB 7.0 software installed on a Laptop with 2.2 GHz Core2Dual CPU is used for signal processing (Fig. 2.1).

Our samples are 20 normal and healthy subjects aged between 20 and 30 with almost similar physical power randomly selected from students of Biomedical Engineering Department, Amirkabir University of Technology, satisfying conditions of having enough sleep and appropriate nutrition, having no considerable physical activity before test, no sedative drug use for at least 24 h before test, with no bone fracture and musculoskeletal disorder close to test, and no pain sensed during tests by subjects. Each individual fills out a form requesting following items:

- Personal information as name, gender, age, height, and weight
- Types of recording signals
- Recording degrees of freedom
- Stimulations and motions
- Processing items requested
- Notes

© The Author(s) 2015   11
B. Mokhlesabadifarahani and V.K. Gunjan, *EMG Signals Characterization in Three States of Contraction by Fuzzy Network and Feature Extraction*, Forensic and Medical Bioinformatics, DOI 10.1007/978-981-287-320-0_2

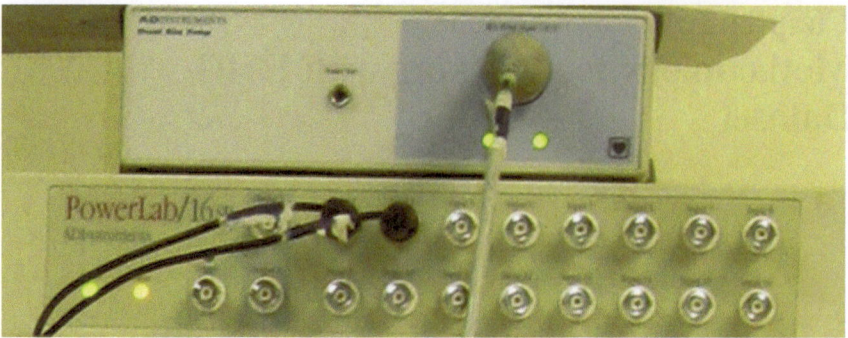

Fig. 2.1 Signal conversion devices PowerLab

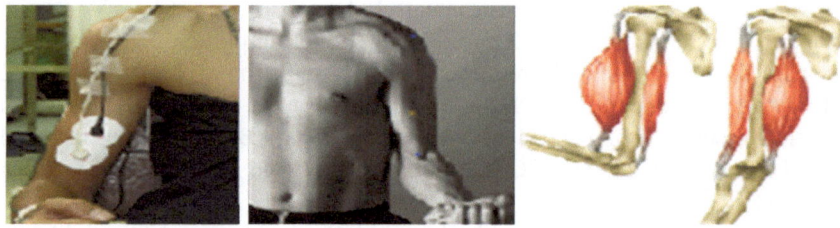

Fig. 2.2 Biceps anatomy and electrode placement

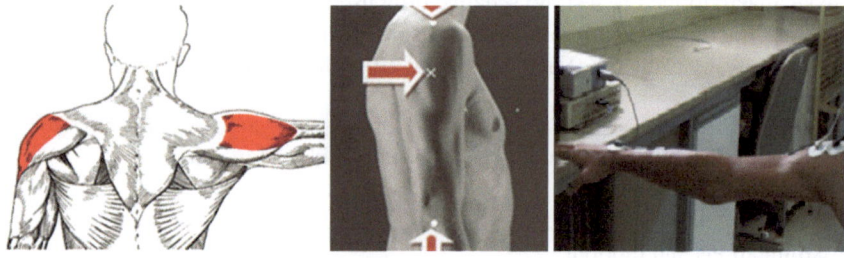

Fig. 2.3 Deltoid anatomy and electrode placement

EMG signals of biceps, deltoid, triceps, tibialis anterior, and quadriceps muscles are recorded in three states of isometric contraction (ISO), maximum voluntary contraction (MVC), and dynamic contractions (Figs. 2.2, 2.3, 2.4, 2.5, and 2.6).

A preprocessing filtering process is then applied to recorded signals. A window consisting of 20,000 samples (10 s) is made cutoff for each signal to be processed and analyzed (Fig. 2.7).

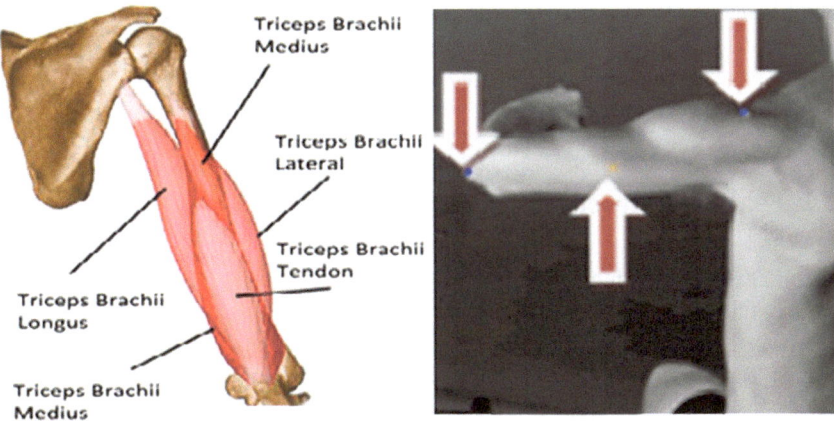

Fig. 2.4 Triceps anatomy and electrode placement

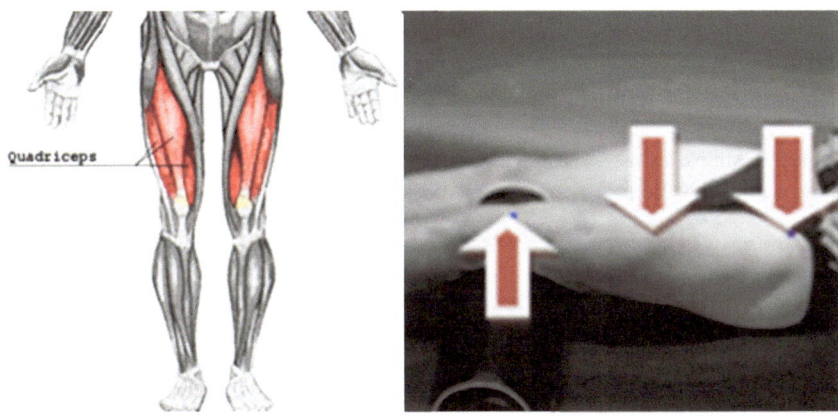

Fig. 2.5 Quadriceps anatomy and electrode placement

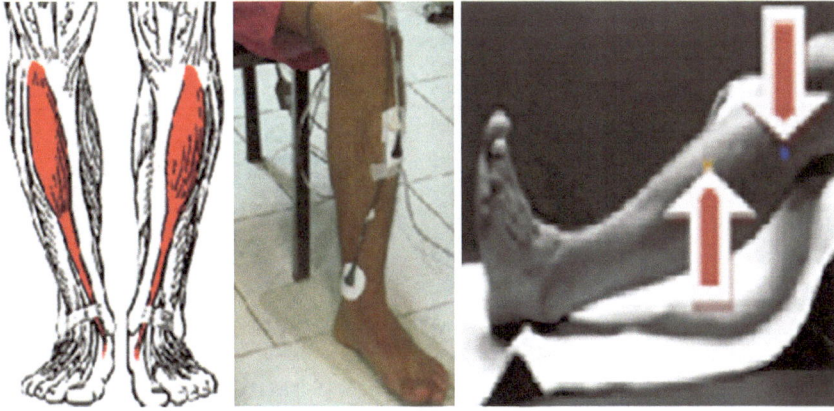

Fig. 2.6 Tibialis anterior anatomy and electrode placement

Fig. 2.7 Illustrated samples
in 10 s

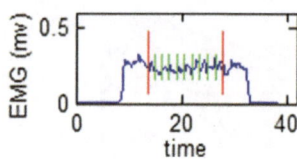

2.2 Feature Extraction

Feature extraction, which is step to measure features or properties from input data, is essential in pattern recognition system design. Goal of feature extraction is to characterize an object to be recognized by measurements whose values are very similar for objects in same category, and very different for objects in different categories. Computational complexity and class discrimination are two main factors for determining best feature set.

A set of features are listed in Table 2.1 along with their descriptions. The primary purpose of this work is to use these features to find an optimum set best describing and characterizing EMG signals. The nonlinear classifier used in this work is a five-layer neuro-fuzzy network; its inputs are selected features among feature list given in Table 2.2. In our experiments, it has been found that there

Table 2.1 Muscle test assessments by professional experts

Case number	Biceps	Triceps	Middle deltoid	Quadriceps	Tibialis anterior
1	5	+5	5	−	5
2	+5	+5	+5	−	+5
3	5	−5	+4	−	−5
4	−5	5	+4	−	−5
5	+4	4	4	−5	−5
6	4	4	+3	+3	−4
7	5	+4	+4	5	−5
8	5	5	5	−	5
9	+4	5	4	+4	5
10	+4	+4	+4	+4	5
11	5	5	+4	5	5
12	+5	+5	+5	−	−
13	+4	+4	+4	5	+4
14	4	+4	+4	5	+4
15	+4	4	4	5	+5
16	−	−	−	−	+4
17	+5	+5	+5	−	+5
18	+4	4	4	+4	+4
19	+5	+5	+5	−	+5
20	+4	+4	4	+4	5
21	+4	5	+4	−	+5

[*]Because of laboratory condition, quadriceps just had been evaluated in female

Table 2.2   Extracted features of recording EMG signals (k: sample number and N: total number of samples).

Item	Feature name	Feature definition	Description
1	Integration of absolute of EMG signal	$\text{IEMG} = \sum_{k=1}^{N} |\text{emg}_k|$	Related to muscle activity
2	Mean absolute value of signal	$\text{MAV} = \frac{1}{N} \sum_{k=1}^{N} |\text{emg}_k|$	Related to muscle contraction points
3	Root mean squared of signal	$\text{RMS} = \left(\sqrt{\sum_{k=1}^{N} \text{emg}_k^2} \right) / N$	Related to muscle contraction indication with constant force before starting the muscle fatigue
4	Wave length of signal	$\text{WL} = \sum_{k=1}^{N-1} |\text{emg}_{k+1} - \text{emg}_k|$	–
5	Difference absolute mean value of signal	$\text{DAMV} = \frac{1}{N-1} \sum_{k=1}^{N-1} |\text{emg}_{k+1} - \text{emg}_k|$	Mean of WL
6	Variance of signal	$\text{VAR} = \frac{1}{N-1} \sum_{k=1}^{N} \text{emg}_k^2$	Related to signal power
7	Zero crossings of signal	$\text{ZC} = \sum_{k=1}^{N} \text{sgn}\left(-\text{emg}_k \text{emg}_{k+1} \right)$ $\text{sgn}(x) = \begin{cases} 1 & \text{if } x > 0 \\ 0 & \text{otherwise} \end{cases}$	For measuring frequency shift and showing number of signal sign varying
8	Wilson amplitude of signal	$\text{WAMP} = \sum_{k=1}^{N} f\left(|\text{emg}_k - \text{emg}_{k+1}| \right)$ $f(x) = \begin{cases} 1 & \text{if } x > \text{threshold} \\ 0 & \text{otherwise} \end{cases}$	An indication of action potential firing of motional unit and therefore muscle contraction level

(continued)

Table 2.2 (continued)

Item	Feature name	Feature definition	Description
9	Cepstrum coefficients of signal	$c_1 = a_1$ $c_n = -\sum_{k=1}^{N}\left(1 - k/N\right)a_k c_{N-k} - a_N$ $CC = [c_1\ c_2 \ldots c_N]$	Containing information about power spectrum of signal and belonging to static features
10	Simple square integral of signal	$SSI = \frac{1}{N}\sum_{k=1}^{N}|emg_k|^2$	An indication of energy of signal
11	Conduction velocity of signal	$CV = \left(\frac{1}{N-1}\sum_{k=1}^{N}emg_k^2\right)$	Similar to RMS feature
12	Mean absolute value slope of signal	$MAVS = MAV(i+1) - MAV(i)$	Related to muscle contraction variations

is a trade-off between classification accuracy and computational complexity. Therefore, for off-line signal processing high classification accuracy is followed through assigning more effective features and for online processing in which computation time is concerned minimum possible number of features should be chosen.

2.3 Neuro-Fuzzy Classifier

These days, neuro-fuzzy systems have been used in broad span of commercial and industrial applications that require analysis of indefinite and indecisive information [45, 47–49]. Hybrid integrated neuro-fuzzy is the major interest of research as it makes use of complementarities' strength of artificial neural network and fuzzy inference systems [45]. ANFIS, a neuro-fuzzy model, is used in this study, which is hybrid technology of integrated neuro-fuzzy model and a part of MATLAB's Fuzzy Logic Toolbox [46]. ANFIS is called a hybrid learning method since it combines gradient descent and least squares method. Gradient descent method is used for premising and tuning parameters that define membership functions. Least squares method is used for identifying parameters that define coefficients of each output equation [46]. The normal structure of ANFIS-like structure used in MATLAB's Fuzzy Logic Toolbox is employed in this research for its efficiency and applicability in clustering and classification problems. The ANFIS has a five-layer structure as described later in this section. Input layer accepts features so that the number of input nodes is same as the number of features. As it was mentioned earlier, ANFIS is composed of gradient descent and least squares methods for learning purpose (Fig. 2.8).

To represent fuzzy inference system, fixed number of layers is presented structurally. ANFIS in comparison with other neuro-fuzzy networks has high training speed, most effective learning algorithm, and simplicity of software [50]. ANFIS is the best function approximator and classifier among neuro-fuzzy models, and its fast convergence is comparable to other neuro-fuzzy models, although it was one of the first integrated hybrid neuro-fuzzy models [51]. Besides, ANFIS affords superior results when applied without any pre-training [52]. Most of the neuro-fuzzy inference systems are based on Takagi–Sugeno or Mamdani type. For model-based applications, Takagi–Sugeno fuzzy inference system is usually used [53, 54]. However, Mamdani fuzzy inference system is used for faster heuristics but with a low performance [55]. High accuracy and easy interpretation of Takagi–Sugeno system makes it a general tool for approximation. The generality of Takagi–Sugeno type is used for identification of complex systems [56]. These systems usually have expensive computations and require complicated learning approaches, but their performance is notable. Typical fuzzy rule for Takagi–Sugeno system is

$$\text{If } x_1 \text{ is } \text{MF}_i^1 \quad \text{and} \quad x_2 \text{ is } \text{MF}_i^2, \quad \text{Then } O \text{ is } \Gamma_i,$$

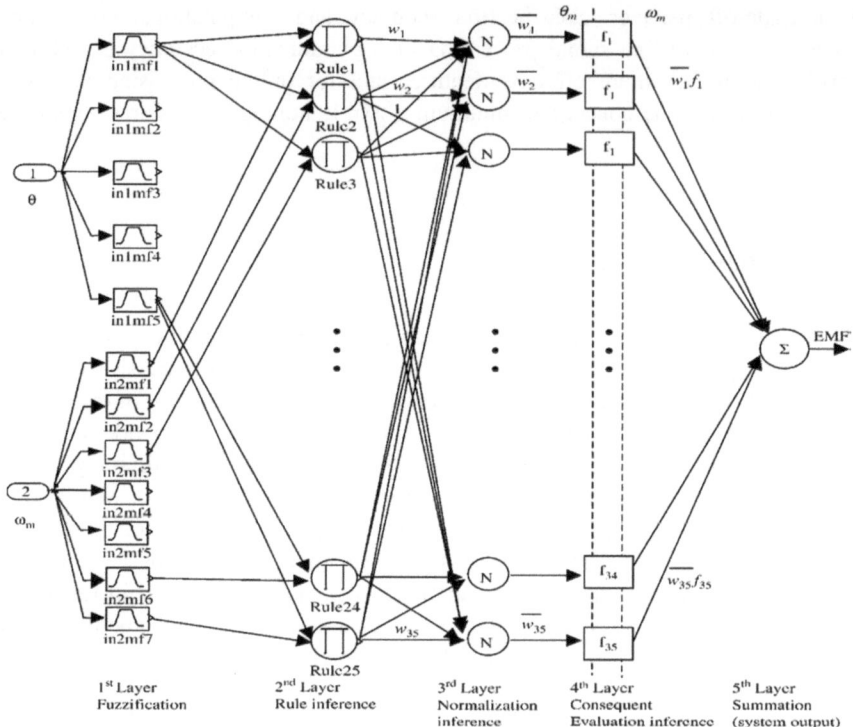

Fig. 2.8 The structure of neuro-fuzzy model

where MF_i^1 and MF_i^2 are fuzzy sets in antecedent and Γ_i is a crisp function in consequent. Usually, function O is a first order or a zero order for Takagi–Sugeno fuzzy inference [57]. In this study, first-order Takagi–Sugeno system is used for fuzzy inference part, and its structure is presented in Fig. 1.

Each one of the five layers of ANFIS performs a specific role for fuzzy inference system as follows:

Layer 1: First layer nodes are adaptive and generate membership grades for input set. Because of their smoothness and concise notation, Gaussian membership function, well known in fields of probability and statistics, is becoming increasingly popular function in fuzzy sets theory. In this study, Gaussian membership functions are used which can be automatically generated by ANFIS of MATLAB. The number of input nodes is the same as the number of features used for classification. Therefore, the number of features specifies the structure of neuro-fuzzy system and also complexity of learning procedure as it directly corresponds to the number of premise parameters as Gaussian membership functions.

Assume input is an *N-D* vector (same of extracted features) and we have 3 membership functions for each input array as

$$X = \begin{bmatrix} x_1 \\ x_2 \\ \vdots \\ x_N \end{bmatrix}_{N \times 1}, \tag{2.1}$$

The Gaussian membership functions are given by

$$\text{GaussMF}_{D-\text{mf}}^{i,j} = \text{Exp}\left(-\frac{\left(x_i - C_j^i\right)^2}{2\delta_j^{i2}}\right), \tag{2.2}$$

where x_i is an input array, and C_j^i and δ_j^i are antecedent parameters expressing centers and standard deviations of Gaussian membership functions of input vector, respectively. Output of this layer is a $M \times N$ matrix with M membership value for each of the N input variables.

With assumption mentioned above, for N input variables and 3 membership functions for each variable, output of first layer is

$$\text{MF}_{D=N,\ \text{mf}=3}^{\text{Gaussian}(X,C,\Sigma)} = \begin{bmatrix} \text{MF}_1^1 & \text{MF}_1^2 & \cdots & \text{MF}_1^N \\ \text{MF}_2^1 & \text{MF}_2^2 & \cdots & \text{MF}_2^N \\ \text{MF}_3^1 & \text{MF}_3^2 & \cdots & \text{MF}_3^N \end{bmatrix}, \tag{2.3}$$

where MF_j^i is the membership value of the ith input to its jth membership function.

Layer 2: All nodes are fixed in this layer. All potential rules between the inputs are formulated applying fuzzy intersection (AND). The operation of product is used to estimate the firing strength of each rule. The output of this layer is a $M^N \times 1$ matrix (N input variables and M membership functions for each input variable), that is,

$$W_{3^N \times 1} = \left[w_{(i=1,2,\ldots,3^N)}^i\right]_{3^N \times 1}, \tag{2.4}$$

Layer 3: In the third layer, the nodes are also fixed nodes. The nodes are symbolized by a notation of N, and the ratio of the ith rule's activation level to the total of all activation levels is computed. The output of this layer denominates as normalized firing strength

$$\overline{W}_{3^N \times 1} = \frac{W_{3^N \times 1}}{\sum_{i=1}^{3^N} w^i}, \tag{2.5}$$

Layer 4: The nodes are adaptive nodes in this layer. Each adaptive node i calculates the contribution of the ith rule toward the overall output as simply product of normalized firing strength and a first-order polynomial (for a first-order Sugeno model). Parameters in this layer are referred to as consequent parameters which shape output of this layer as

$$O^i = \Gamma^i \cdot \overline{w}^i \tag{2.6}$$

Layer 5: There is only one single fixed node in the fifth layer which calculates overall output as summation of contribution from each rule

$$O_{mem.value} = \Gamma_{1 \times 3^N} \overline{W}_{3^N \times 1}, \tag{2.7}$$

where

$$\Gamma_{1 \times 9} = \begin{bmatrix} x_1 & x_2 & 1 \end{bmatrix} Coeff_{3 \times 3^N}, \tag{2.8}$$

That $Coeff_{3 \times 3^N}$ is a matrix of consequent parameters of ANFIS used. It can be observed that there are two adaptive layers in this ANFIS structure, the first and fourth layers. There are two modifiable matrices of parameters $C_{3 \times N}$ and $\Sigma_{3 \times N}$ which shape input Gaussian membership functions. These parameters are so-called premise parameters. In the fourth layer, there is also a modifiable matrix of parameters $Coeff_{3 \times 3^N}$, pertaining to first-order polynomial. These parameters are so-called consequent parameters [58, 59].

Both premise and consequent matrices of parameters are adjusted during learning procedure aiming to make ANFIS output match training data. Least squares method can be used to identify optimal values of these parameters easily. When premise parameters are not fixed, search space becomes larger and convergence of training becomes slower. A hybrid algorithm combining least squares method and gradient descent method is adopted to solve this problem. Hybrid algorithm is composed of a forward pass and a backward pass. Least squares method (forward pass) is used to optimize consequent parameters with premise parameters fixed. Once optimal consequent parameters are found, backward pass starts immediately. Gradient descent method (backward pass) is used to adjust optimally premise parameters corresponding to fuzzy sets in input domain. Output of ANFIS is calculated by employing consequent parameters found in forward pass. Output error is used to adapt premise parameters by means of a standard back-propagation algorithm. It has been proven that this hybrid algorithm is highly efficient in training ANFIS [58, 59]. Once ANFIS is structured and learned, parameters are deterministic and classification can be executed.

Chapter 3
Results

Flowchart of EMG signal processing executed in our experiments is shown in Fig. 3.1. According to flowchart, after data acquisition phase, all recorded signals underwent noise filtering as preprocessing phase. Long-time recording signals are cut off in a windowing procedure as long as 20,000 samples or 10 s records with 2 kHz sampling rate. Each window is split into sub-windows with length of 100–5,000 samples. For all windows, 70 % of samples are set for training procedure and rest of samples for testing purpose. After multiple runs of training and testing procedures for different lengths of windows, windows length of 2,000 samples (corresponding to 1 s signal recording) was chosen. Therefore, each window (with 20,000 samples) is split into 10 sub-windows each one with 2,000 samples. Seventy percent of sub-windows are still considered for training purpose (7 sub-windows) and rests for testing purpose (3 sub-windows).

For evaluating classifier, mean-squared error is used which is most common criterion defined as below:

$$\text{MSE} = \frac{1}{N} \sum_{i=1}^{N} (y_i - \bar{y}_i)$$

where y_i and $\bar{y}_i$ are real and desired outputs of network, respectively, and N is total number of samples. MSE of training process shows trainability of system, and MSE of testing samples indicates system's modeling capability.

True classification rate is defined as rate of true assigned samples to their classes to whole number of samples as below:

$$\text{Classification Rate} = \frac{\text{True assigned samples}}{\text{Total number of samples}}$$

© The Author(s) 2015

B. Mokhlesabadifarahani and V.K. Gunjan, *EMG Signals Characterization in Three States of Contraction by Fuzzy Network and Feature Extraction*, Forensic and Medical Bioinformatics, DOI 10.1007/978-981-287-320-0_3

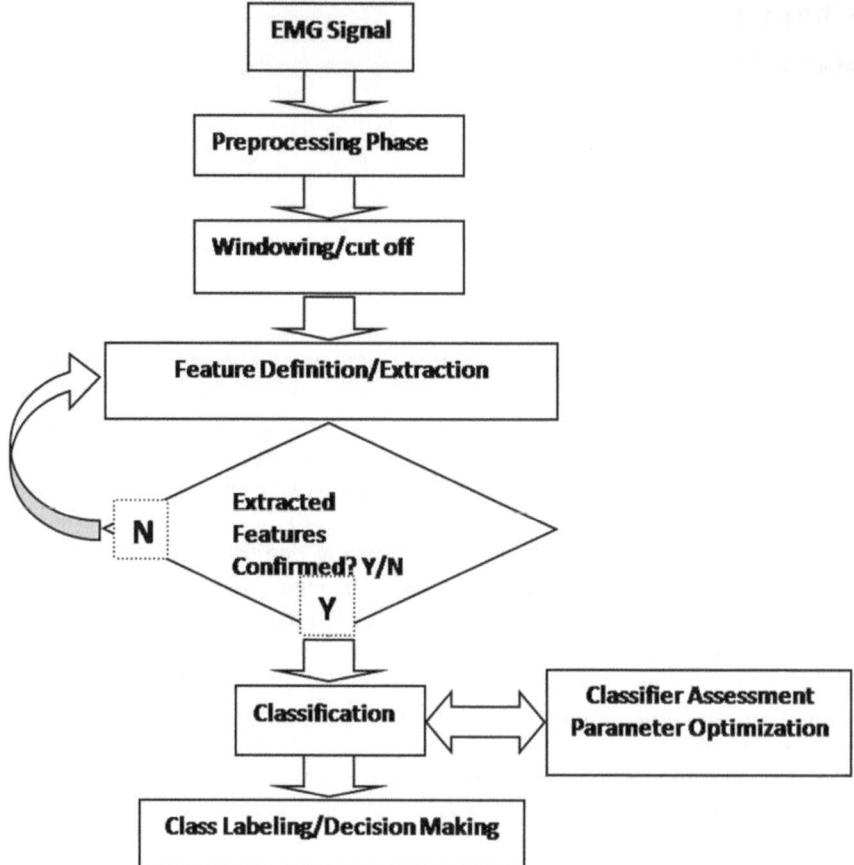

Fig. 3.1 Flowchart of EMG signal characterization process used in this work

The flowchart of procedure sequence of EMG signal characterization is shown in Fig. 3.1. As it is illustrated in flowchart, there are two main assessment sections on process, one for evaluating extracted features and another one for evaluating and structure optimization of classifier. Therefore, evaluation of potential features is taken into account. The reference classifier for this step is a multilayer perceptron (MLP) as an efficient artificial neural network with least square back propagation learning algorithm.

The MLP used in this part of simulation has 1 hidden layer and 20 neurons in the hidden layer. Transfer function was used as tangent sigmoid mathematics function. Inputs of MLP are extracted features, and network is trained based on training samples described earlier.

According to Table 3.1, combinations of RMS + WL and RMS + MAV + WL yield best result for quadriceps muscle. For biceps muscle, lowest MSE corresponds to combination of MAV and VAR according to Table 3.2.

Table 3.1 Training and testing errors of MLP network for one or a set of features corresponding to the EMG signals of quadriceps muscle contraction

Item	Extracted feature(s)	MSE of training	MSE of testing
1	RMS	0.0642	0.1183
2	MAV	0.0562	0.0894
3	ZC	0.0754	0.0644
4	SSI	0.0650	0.0927
5	WL	0.0614	0.0854
6	RMS + MAV	0.0532	0.0721
7	RMS + WL	0.05	0.0609
8	RMS + VAR	0.0648	0.0847
9	RMS + SSI	0.0625	0.0858
10	RMS + ZC	0.0622	0.0751
11	MAV + VAR	0.0537	0.0741
12	RMS + MAV + WL	0.0498	0.0638
13	RMS + MAV + ZC	0.0593	0.0654
14	RMS + MAV + VAR	0.0571	0.0709

Table 3.2 Training and testing errors of the MLP network for one or a set of features corresponding to the EMG signals of biceps muscle in a dynamic test

Item	Extracted feature(s)	MSE of training	MSE of testing
1	IEMG	0.4259	0.0510
2	MAV	0.4239	0.0982
3	SSI	0.5844	0.0360
4	RMS	0.6869	0.0357
5	WL	0.5487	0.0476
6	ZC	0.4884	0.0585
7	CV	0.6820	0.1095
8	VAR	0.7064	0.0138
9	RMS + MAV	0.6415	0.0159
10	RMS + SSI	0.6202	0.0345
11	RMS + VAR	0.6649	0.0104
12	RMS + ZC	0.5940	0.0351
13	RMS + WL	0.6609	0.0128
14	MAV + VAR	0.6407	0.0088

From results of Tables 3.3, 3.4, and 3.5, it can be inferred that isometric contraction test, compared to two other contractions, has lower MSE values when it is modeled. It should be noted that these MSE errors correspond to distance to relative class and misclassification of force of a muscle with a higher or lower class of force do not affect course of treatment significantly. In Tables 3.3, 3.4, and 3.5, more appropriate features are colored red according to their MSE values which are selective features for isometric contraction (ISO), maximum voluntary

24   3 Results

Table 3.3 Best results for biceps, deltoid, triceps, quadriceps, and tibialis anterior muscles (ISO)

Muscle	Gender	Appropriate features	MSE of training	MSE of testing
Biceps	Female	RMS	0.0509	0.0566
	Female	MAV + SSI	0.0565	0.0452
	Female	MAV + WL	0.0585	0.0399
	Male	RMS	0.0394	0.0229
	Male	RMS + WL	0.0396	0.0222
Deltoid	Male	RMS + WL	0.1585	0.1060
	Male	RMS + SSI + VAR + WL + IEMG	0.1921	0.0672
Triceps	Female	RMS + WL	0.0403	0.0457
	Female	RMS + WL + MAV	0.0522	0.0442
	Male	RMS + ZC	0.0161	0.0095
	Male	RMS + MAV + WL	0.013	0.0076
	Male	RMS + MAV + ZC	0.0127	0.0062
Quadriceps	Female	RMS + WL	0.065	0.079
	Female	RMS + ZC	0.0668	0.0799
	Female	RMS + MAV + VAR	0.0906	0.0792
Tibialis anterior	Female	RMS + WL	0.0368	0.0453
	Female	RMS + VAR	0.0387	0.0436
	Male	RMS + WL	0.0893	0.0769
	Male	RMS + ZC	0.0959	0.1849

Table 3.4 Best results for biceps, deltoid, triceps, quadriceps, and tibialis anterior muscles (MVC)

Muscle	Gender	Appropriate features	MSE of training	MSE of testing
Biceps	Female	RMS + SSI	0.0306	0.0138
	Female	RMS + ZC	0.029	0.0206
	Male	RMS + ZC + MAV	0.0208	0.0165
Deltoid	Female	ZC	0.0255	0.0217
	Male	RMS	0.1651	0.1619
Triceps	Female	RMS + MAV + ZC	0.0624	0.0596
	Female	RMS + MAV + WL	0.067	0.0583
	Male	MAV	0.0109	0.1132
	Male	WL	0.0754	0.1112
Quadriceps	Female	RMS + MAV	0.577	0.0661
	Female	RMS + WL	0.0617	0.0668
Tibialis anterior	Female	RMS + WL	0.0424	0.0334
	Female	RMS + MAV	0.0484	0.0341
	Male	RMS + WL	0.0451	0.0405
	Male	RMS + WL + MAV	0.0549	0.0605

Table 3.5 Best results for biceps, deltoid, triceps, quadriceps, and tibialis anterior muscles (dynamics)

Muscle	Gender	Appropriate features	MSE of training	MSE of testing
Biceps	Female	IEMG	0.0510	0.0425
	Female	MAV	0.0982	0.0423
	Male	ZC	0.1556	0.1083
Deltoid	Female	RMS + WL	0.0229	0.0194
	Male	RMS + ZC	0.0881	0.1565
Triceps	Female	RMS + MAV	0.1054	0.0981
	Female	RMS + VAR	0.1042	0.0927
	Male	ZC	0.0425	0.0517
	Male	RMS + MAV	0.0852	0.0861
Quadriceps	Female	RMS + WL	0.05	0.609
	Female	RMS + WL + MAV	0.0498	0.639
Tibialis anterior	Female	RMS + WL	0.0851	0.085
	Female	RMS + ZC	0.0722	0.053
	Male	RMS	0.1017	0.1409
	Male	RMS + WL	0.0983	0.103

contraction (MVC), and dynamic contraction, respectively. One primary purpose of developing an expert system to perform feature extraction and sample classification is to achieve a standard evaluation procedure and to reduce therapist affectivity on evaluation quality.

After most suitable features are extracted by reference classifier, an evaluation is performed to find a robust and efficient classifier. Neuro-fuzzy network is a potential for this task which is compared to some other well-known classifiers in this part of experiment.

For neuro-fuzzy classifier, a combination of least squares and back propagation method was used as learning algorithm. Trapezoidal and Gaussian membership functions are commonly used as shape of fuzzy sets of inputting nodes. Number of 2–4 membership functions is suggested for each variable in EMG signal modeling problem.

Table 3.6 shows results of implementing five types of classifiers for classification of EMG signals according to extracted features. Due to difference between muscles power of two groups of gender, male and female, we separated males and females in analysis of their EMG signals of mentioned muscles. Classes of muscle forces are separated and samples are assigned to their relative classed. True assignments of samples to their classes define classification rate in percent as criteria for evaluation of classifier in addition to training and testing capability.

Table 3.6 Classifier evaluation for classification of EMG signals (muscle force) according to extracted features

Classifier	Features	MSE of train	MSE of test	Run time (s)	Uncertainty (%)	Classification rate (%)
K-NN	RMS + WL	–	–	~2	0	~76
FFNN-1	RMS + WL	0.024	0.021	~12	~4	~79
FFNN-2	RMS + WL	0.022	0.020	~18	~7	~83
FFNN-3	RMS + WL	0.017	0.018	~22	~9	~85
ERNN	RMS + WL	~5e−06	0.312	~14	~12	~71
F.C-means	RMS + WL	–	–	~5	~5	~80
NFS-1	RMS + WL	0.003	0.011	~7	~2	~82
NFS-2	RMS + WL	0.001	0.002	~11	~2	~85
NFS-3	RMS + WL	3.1e−04	4.7e−04	~23	~3	~87
NFS-4	RMS + WL	2.3e−04	3.9e−04	~31	~3	~88
NFS-5	RMS + WL	2.1e−04	3.3e−04	~39	~3	~89
NFS-6	RMS + WL	2.1e−04	2.9e−04	~43	~4	~91

K-NN K-nearest neighbor, *FFNN* feed-forward neural network, *ERBNN* Elman recurrent neural network, *F.C-means* fuzzy C-means, *NFS* neuro-fuzzy system

Except deterministic algorithm of K-NN, rests involve an uncertainty which means respective variations of outputs in a sequence of executions

FFNN-1: ([10, 1], 'Logsig', 'Purelin', 'Trainlm', 500)
FFNN-2: ([20, 1], 'Logsig', 'Purelin', 'Trainlm', 500)
FFNN-3: ([20, 1], 'Logsig', 'Purelin', 'Trainlm', 1000)
ERNN: (Spread = 0.01)
NFS-1: (NumMFs = 2, MFtype: 'Gaussmf', 300)
NFS-2: (NumMFs = 2, MFtype: 'Gaussmf', 500)
NFS-3: (NumMFs = 3, MFtype: 'Gaussmf', 300)
NFS-4: (NumMFs = 3, MFtype: 'Gaussmf', 500)
NFS-5: (NumMFs = 4, MFtype: 'Gaussmf', 300)
NFS-6: (NumMFs = 4, MFtype: 'Gaussmf', 500)

Chapter 4
Conclusions and Inferences of Present Study

Feed-forward neural network with three structural parameters, Elman recurrent neural network, fuzzy C-means, and 6 structures of neuro-fuzzy system are employed to first model samples and then to assign them to their respective classes. It can be seen for offline processing of EMG signals where accuracy and adaptability are followed by classifier, neuro-fuzzy system outperforms other tools given in Table 3.8. For real-time applications, timing characteristics of neuro-fuzzy systems should be taken into account since training this type of classifier requires a training procedure and training procedure takes longer times for larger number of fuzzy rules. Nevertheless, whenever working in offline mode and when accuracy is only concerned, neuro-fuzzy system yields satisfactory results comparatively. EMG signals of biceps, deltoid, triceps, tibialis anterior, and quadriceps muscles were recorded in three states of isometric contraction (ISO), maximum voluntary contraction (MVC), and dynamic contractions from 20 normal subjects aged between 20 and 30; half of them are male. Totally, 14 extracted features are analyzed to find which of them or which set of them is discriminative and selective for muscle force classification. Neuro-fuzzy system is trained with 70 % of recorded EMG cutoff windows, and then, it is employed for classification and modeling purposes. For each muscle, most effective extracted features are found for males and females separately by a reference classifier. Finally, neuro-fuzzy classifier is validated in comparison with some other well-known classifiers for classification of recorded EMG signals with three states of contractions corresponding to extracted features. It was found that combinations of RMS + WL and RMS + MAV + WL yield best result for quadriceps muscle, and for biceps muscle, lowest MSE corresponds to combination of MAV and VAR features. It was also inferred that isometric contraction test, compared to two other contractions, has lower MSE values when it is modeled so it is more discriminative and effective in sample classifications.

© The Author(s) 2015
B. Mokhlesabadifarahani and V.K. Gunjan, *EMG Signals Characterization in Three States of Contraction by Fuzzy Network and Feature Extraction*, Forensic and Medical Bioinformatics, DOI 10.1007/978-981-287-320-0_4

Appendix

See Figs. A.1, A.2, A.3 and A.4.

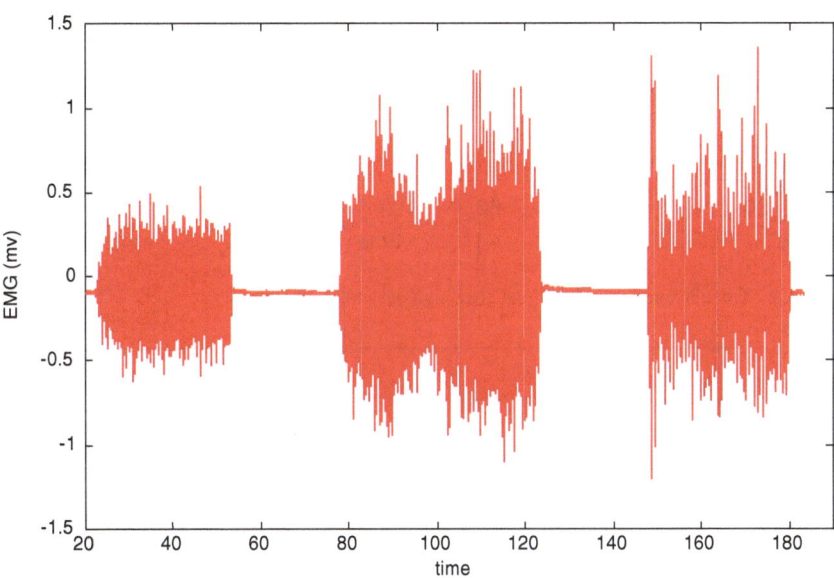

Fig. A.1 Illustrate EMG of tibialis anterior

B. Mokhlesabadifarahani and V.K. Gunjan, *EMG Signals Characterization
in Three States of Contraction by Fuzzy Network and Feature Extraction*,
Forensic and Medical Bioinformatics, DOI 10.1007/978-981-287-320-0

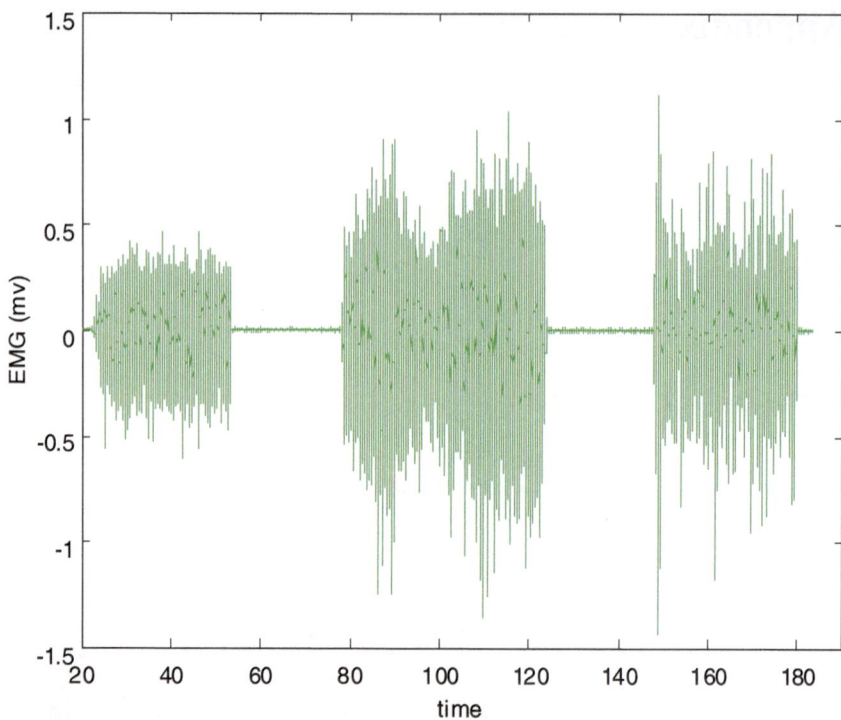

Fig. A.2 Tibialis anterior EMG signal after high pass filter 10 Hz

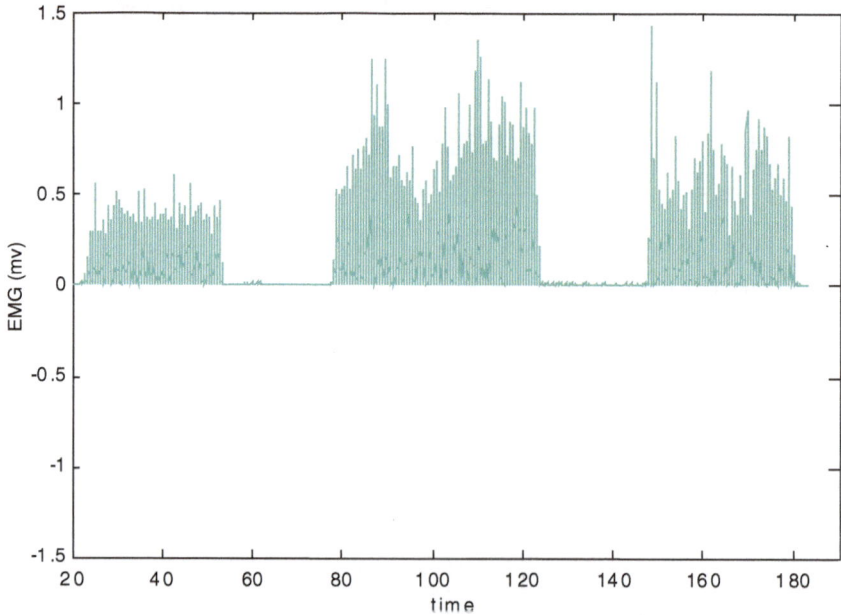

Fig. A.3 EMG signal after absolute value

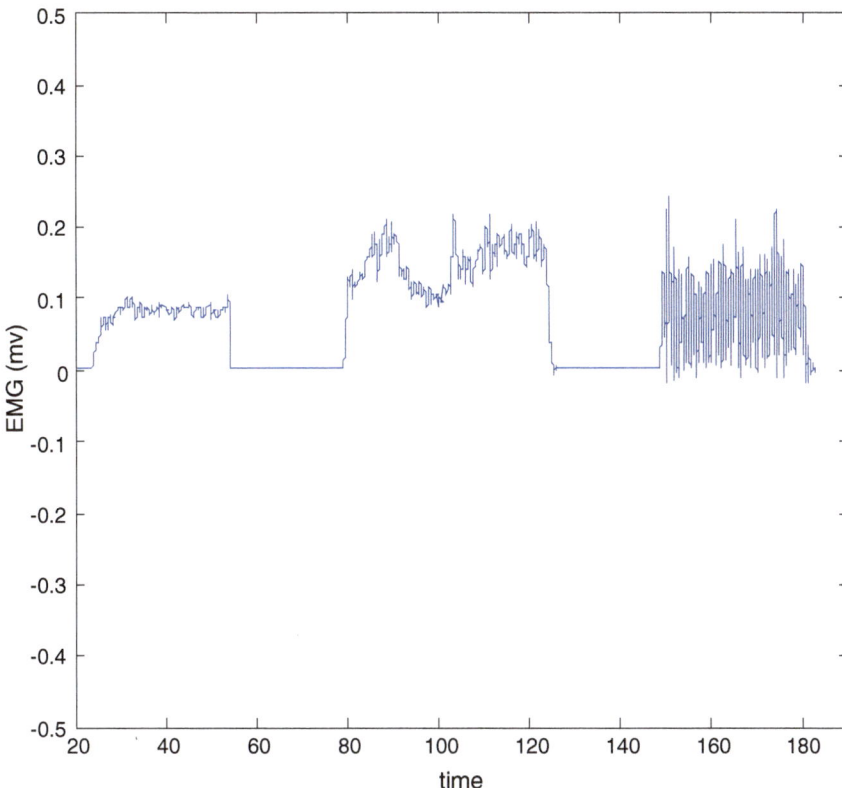

Fig. A.4 Muscle force extraction after filtering with low pass filter 2 Hz

References

1. Katsis CD, Exarchos TP, Papaloukas C, Goletsis Y, Fotiadis DI, Sarmas I (2007) A two-stage method for MUAP classification based on EMG decomposition. Comput Biol Med 37:1232
2. Oskoei MA, Hu H (2007) Myoelectric control systems—a survey. Biomed Signal Process Control 2:275
3. Clancy EA, Morin EL, Merletti R (2002) Sampling noise reduction and amplitude estimation issues in surface electromyography. J Electromyogr Kinesiol 12:1
4. Zardoshti-Kermani M, Wheeler BC, Badie K, Hashemi RM (1995) EMG feature evaluation for movement control of upper extremity prostheses. IEEE Trans Rehab Eng 3:324
5. Boostani R, Moradi MH (2003) Evaluation of forearm EMG signal features for control of a prosthetic hand. Physiol Meas 24:309
6. Phinyomark A, Limsakul C, Phukpattaranont P (2009) A novel feature extraction for robust EMG pattern recognition. J Comput 1:71
7. Oskoei MA, Hu H (2008) Support vector machine based classification scheme for myoelectric control applied to upper limb. IEEE Trans Biomed Eng 55:1956
8. Zecca M, Micera S, Carrozza MC, Dario P (2002) Control of multifunctional prosthetic hands by processing electromyographic signal. Crit Rev Biomed Eng 30:459
9. Phinyomark A, Pornchai P, Chusak L (2012) Feature reduction and selection for EMG signal classification. Expert Syst Appl 39:7420
10. Du YC, Lin CH, Shyu LY, Tainsong C (2010) Portable hand motion classifier for multi-channel surface electromyography recognition using grey relational analysis. Expert Syst Appl 37:4283
11. Farfán FD, Politti JC, Felice CJ (2010) Evaluation of EMG processing techniques using information theory. Biomed Eng Online 9:72
12. Khezri M, Jahed M (2009) An exploratory study to design a novel hand movement identification system. Comput Biol Med 39:433
13. Khushaba RN, Al-Ani A, Al-Jumaily A (2010) Orthogonal fuzzy neighborhood discriminant analysis for multifunction myoelectric hand control. IEEE Trans Biomed Eng 57:1410
14. Kim KS, Choi HH, Moon CS, Mun CW (2011) Comparison of k-nearest neighbor, quadratic discriminant and linear discriminant analysis in classification of electromyogram signals based on wrist-motion directions. Curr Appl Phys 11:740
15. Li G, Li Y, Yu L, Geng Y (2011) Conditioning and sampling issues of EMG signals in motion recognition of multifunctional myoelectric prostheses. Ann Biomed Eng 39:1779
16. Li G, Schultz AE, Kuiken TA (2010) Quantifying pattern recognition-based myoelectric control of multifunctional transradial prostheses. IEEE Trans Neural Syst Rehabil Eng 18:185
17. Tenore FVG, Ramos A, Fahmy A, Acharya S, Etienne-Cummings R, Thakor NV (2009) Decoding of individuated finger movements using surface electromyography. IEEE Trans Biomed Eng 56:1427

© The Author(s) 2015
B. Mokhlesabadifarahani and V.K. Gunjan, *EMG Signals Characterization in Three States of Contraction by Fuzzy Network and Feature Extraction*, Forensic and Medical Bioinformatics, DOI 10.1007/978-981-287-320-0

18. Graupe J, Cine K (1975) Functional separation of EMG signals via ARMA identification method for prosthesis control purpose. IEEE Trans Syst Man Cybern 5:252

19. Kelly M, Parker P (1990) Application of neural network to myoelectric signal analysis: preliminary study. IEEE Trans Biomed Eng 37:221

20. Saridis G, Gootee T (1983) EMG pattern analysis and classification for a prosthetic ARMA. IEEE Trans Biomed Eng 30:18

21. Chang G, Kang W, Luh J, Cheng C (1994) Principal component for classification of pre-shaping movement. In: IEEE 17th annual conference, pp 1019–1025

22. Jang GC, Cheng CK, Lai JS, Kuo TS (1994) Using time-frequency analysis technique in classification of surface EMG signals. In: Proceedings of the IEEE 16th annual international conference of the engineering in medicine and biology society, engineering advances: new opportunities for biomedical engineers, pp 1242–1243

23. Wellig P, Moschytz GS (1999) Classification of time-varying signals using time-frequency atoms. Proc First Joint BMES/EMBS Conf 2:953

24. Liyu C, Zhizhong W, Haihong Z (1999) An EMG classification methods on wavelet transform. In: IEEE Proceedings 1st joint BMES/EMBS conference serving humanity, p 565

25. Abel R, Arikidis NA, Forster A (1998) Inter-scale local maxima—a new technique for wavelet analysis for EMG signals. In: Proceedings 20th annual international conference of the IEEE engineering in medicine and biology society, pp 1471–1473

26. Englehart K, Hudgins JL, Parker P, Stevenson G (1999) Improving myoelectric signal classification using wavelet packets and principal components analysis. In: Proceedings of the 1st joint BMES/EMBS conference serving humanity, IEEE, p 569

27. Christodoulou CI, Pattichis CS (1999) Unsupervised pattern recognition for the classification of EMG signals. IEEE Trans Biomed Eng 46:169

28. Tsenov G, Zeghbib AH, Palis F, Shoylev N, Mladenov V (2006) Neural networks for online classification of hand and finger movements using surface EMG signals. In: Proceedings IEEE 8th seminar on neural network applications in electrical engineering, NEUREL, pp 167–171

29. Yoshikawa M, Mikawa M, Tanaka K (2007) A myoelectric interface for robotic hand control using support vector machine. In: IEEE/RSJ international conference on intelligent robots and systems, IROS, pp 2723–2728

30. Momen K, Krishnan S, Chau T (2007) Real-time classification of forearm electromyographic signals corresponding to user-selected intentional movements for multifunction prosthesis control. IEEE Trans Neural Syst Rehabil Eng 15:535

31. Hudgins B, Parker P, Scott R (1993) A new strategy for multifunction myoelectric control. IEEE Trans Biomed Eng 40:82

32. Englehart K, Hudgins B, Parker PA (2001) A wavelet-based continuous classification scheme for multifunction myoelectric control. IEEE Trans Biomed Eng 48:302

33. Englehart K, Hudgins B (2003) A robust, real-time control scheme for multifunction myoelectric control. IEEE Trans Biomed Eng 50:848

34. Chaiyaratana N, Zalzala AMS, Datta D (1996) Myoelectric signals pattern recognition for intelligent functional operation of upper-limb prosthesis. In: Proceedings of the 1st European conference on disability, virtual reality and associated technologies, UK, pp 151–160

35. Au TC, Kirsch RF (2000) EMG-based prediction of shoulder and elbow kinematics in able-bodied and spinal cord injured individuals. IEEE Trans Rehabil Eng 8:471

36. Ajiboye B, Weir RF (2005) A heuristic fuzzy logic approach to EMG pattern recognition for multifunctional prosthesis control. IEEE Trans Neural Syst Rehabil Eng 13:280

37. Chan FHY, Yang YS, Lam FK, Zhang YT, Parker PA (2000) Fuzzy EMG classification for prosthesis control. IEEE Trans Rehabil Eng 8:305

38. Kiguchi K, Tanaka T, Watanabe K, Fukuda T (2003) Design and control of an exoskeleton system for human upper-limb motion assist. In: Proceedings IEEE/ASME international conference on advanced intelligent mechatronics (AIM), pp 926–931

39. Vuskovic M, Du SJ (2002) Classification of prehensile EMG patterns with simplified fuzzy ARTMAP networks. In: Proceedings international joint conference on neural networks, pp 2539–2544

40. Han JS, Bien ZZ, Kim DJ, Lee HE, Kim JS (2003) Human–machine interface for wheelchair control with EMG and its evaluation. In: Proceedings 25th IEEE international conference engineering in medicine and biology society, Mexico, pp 1602–1605

41. Huang Y, Englehart K, Hudgins B, Chan ADC (2005) A Gaussian mixture model based classification scheme for myoelectric control of powered upper limb prostheses. IEEE Trans Biomed Eng 52:1801

42. Fukuda O, Tsuji T, Kaneko M, Otsuka A (2003) A human-assisting manipulator teleoperated by EMG signals and arm motions. IEEE Trans Robot Autom 19:210

43. Chan DC, Englehart K (2003) Continuous classification of myoelectric signals for powered prostheses using Gaussian mixture models. In: Proceedings 25th annual international conference of the IEEE engineering in medicine and biology society, Mexico, pp 2841–2844

44. Chan C, Englehart K (2005) Continuous myoelectric control for powered prostheses using hidden markov models. IEEE Trans Biomed Eng 52:121

45. Akcayol MA (2004) Application of adaptive neuro-fuzzy controller for SRM. Adv Eng Soft 35:129

46. MATLAB (1999) Users guide: fuzzy logic toolbox. The Mathworks Inc., Nantick

47. Chen MY, Linkens DA (2006) A systematic neuro-fuzzy modeling framework with application to material property prediction. IEEE Trans Syst Man Cybern B 31:781

48. Kruse R, Nauck D, Nurnberger A, Merz L (2007) A neuro-fuzzy development tool for fuzzy controllers under MATLAB/SIMULINK. In: Proceedings of the fifth European congress on intelligent techniques and soft computing EUFIT'97, pp 1029–1033

49. Studer L, Masulli F (2007) Building a neuro-fuzzy system to efficiently forecast chaotic time series. Nucl Instr Meth A 389:264

50. Jang JSR, Sun CT, Mizutani E (1997) Neuro-fuzzy soft computing. Prentice-Hall, Englewood Cliffs

51. Akcayol MA (2004) Application of adaptive neuro-fuzzy controller for SRM. Adv Eng Soft 35:129

52. Altug S, Chow MY, Trussell HJ (2004) Fuzzy inference systems implemented on neural architectures for motor. IEEE Trans Industr Electron 46:1069

53. Sugeno M, Tanaka K (1991) Successive identification of a fuzzy model and its applications to predictions of complex systems. Fuzz Set Syst 42:315

54. Sugeno M, Yasukawa T (1993) A Fuzzy-logic-based approach to qualitative modeling. IEEE Trans Fuzz Syst 1:7

55. Thiesing M, Vornberger O (1997) Sales forecasting using neural networks. Int Conf Neural Netw 4:2125

56. Wang L, Langari R (1996) Complex systems modeling via fuzzy logic. IEEE Trans Syst Man Cybern B 26:100

57. Wang WP, Chen Z (2007) A neuro-fuzzy based forecasting approach for rush order control applications. Expert Syst App 35:223

58. Guler I, Ubeyli DE (2005) Adaptive neuro-fuzzy inference system for classification of EEG signals using wavelet coefficients. J Neurosci Meth 148:113

59. Jang JSR (1992) Self-learning fuzzy controllers based on temporal back propagation. IEEE Trans Neural Netw 3:714